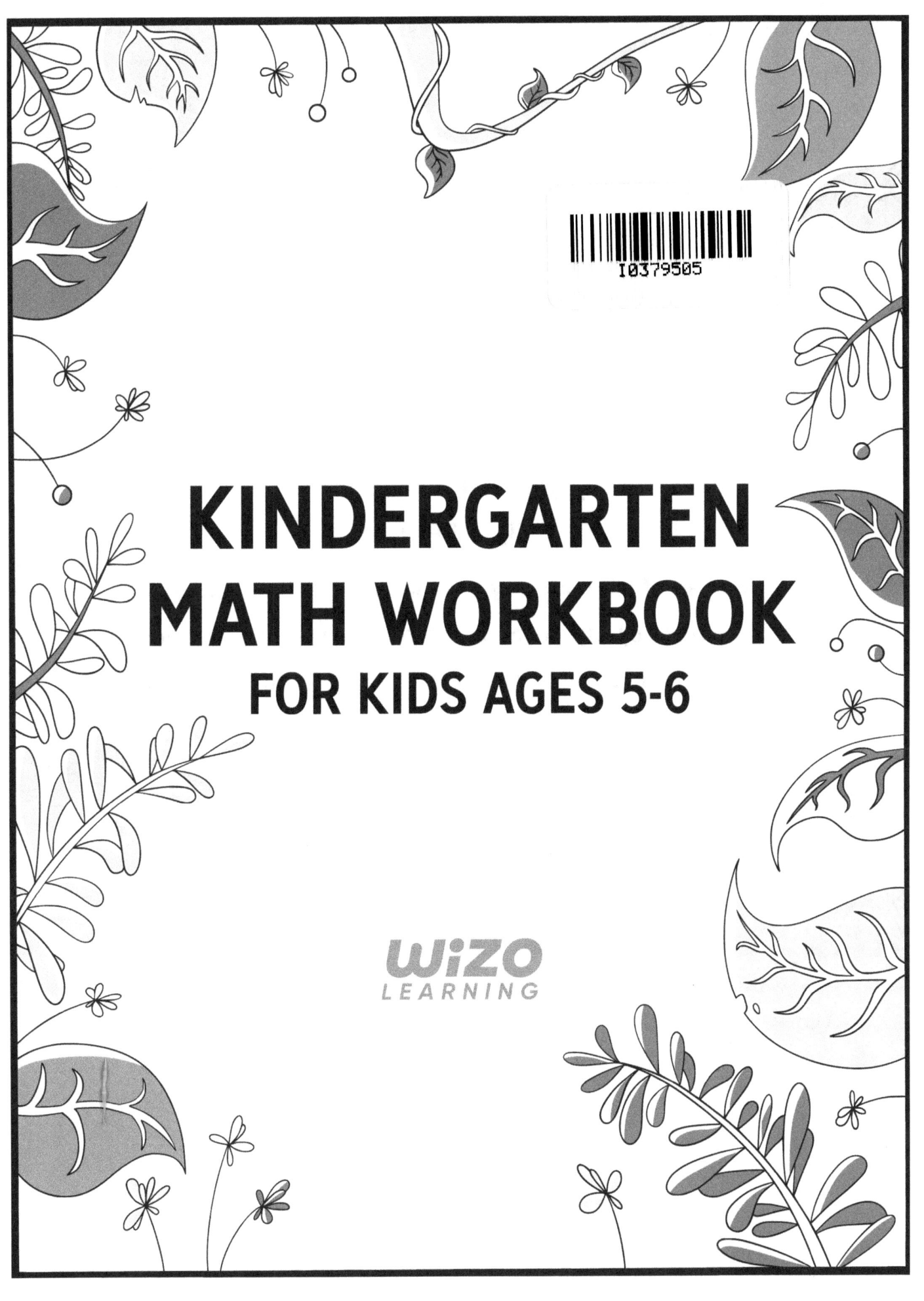

Please consider writing a review!
Just visit: wizolearning.com/review

Copyright 2020. Wizo Learning.
All Rights Reserved.

No part of this book may be reproduced or transmitted in any form or by any means, electronic or mechanical, including photocopying, recording or by any other form without written permission from the publisher.

Have questions? We want to hear from you!
Email us at: support@wizolearning.com

ISBN: 978-1-951806-29-3

# FREE BONUS

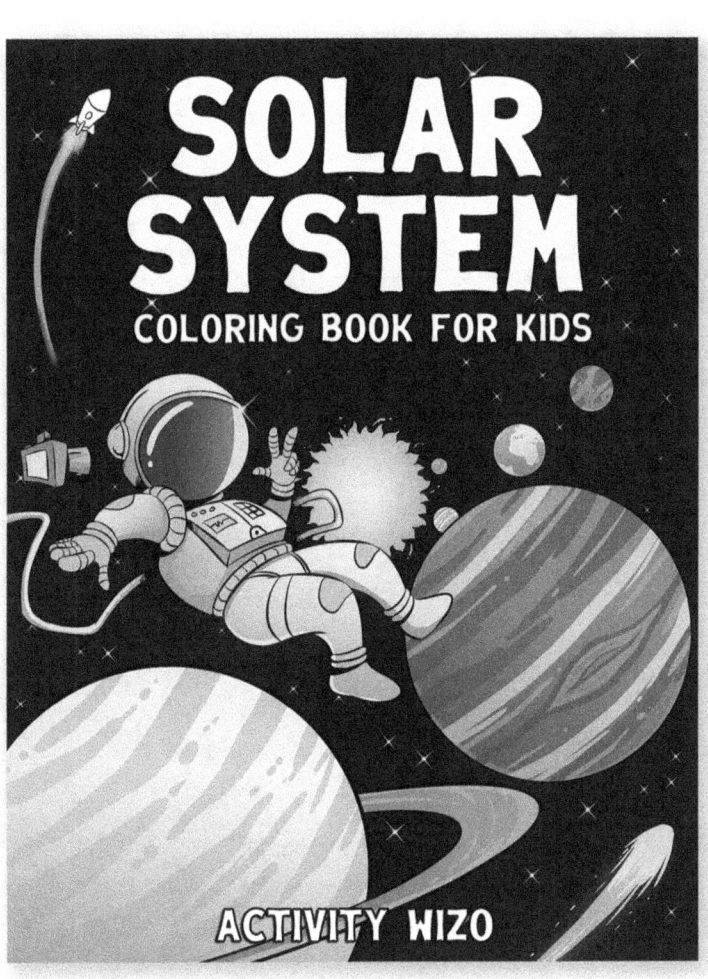

Just visit:
wizolearning.com/solar

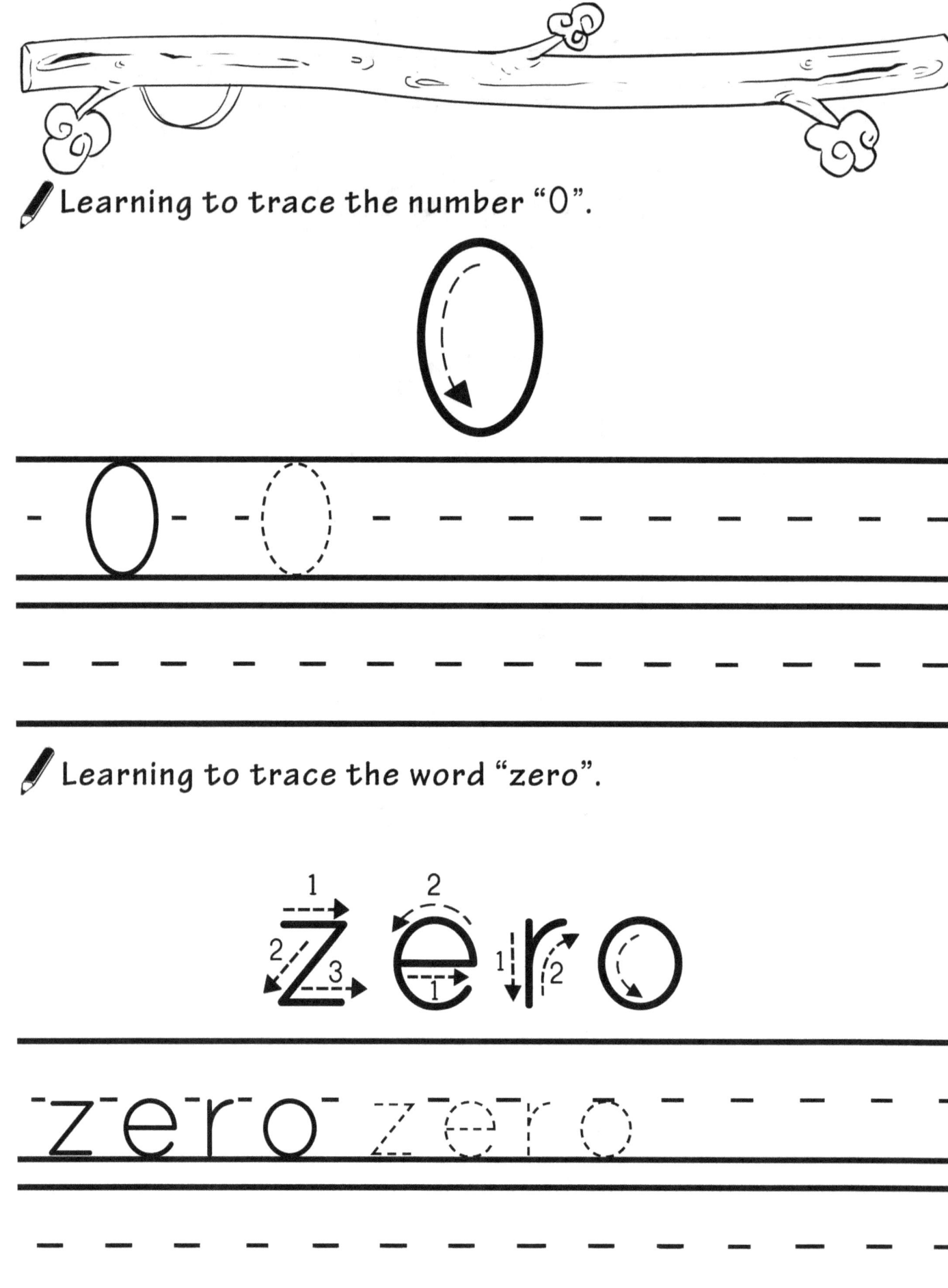

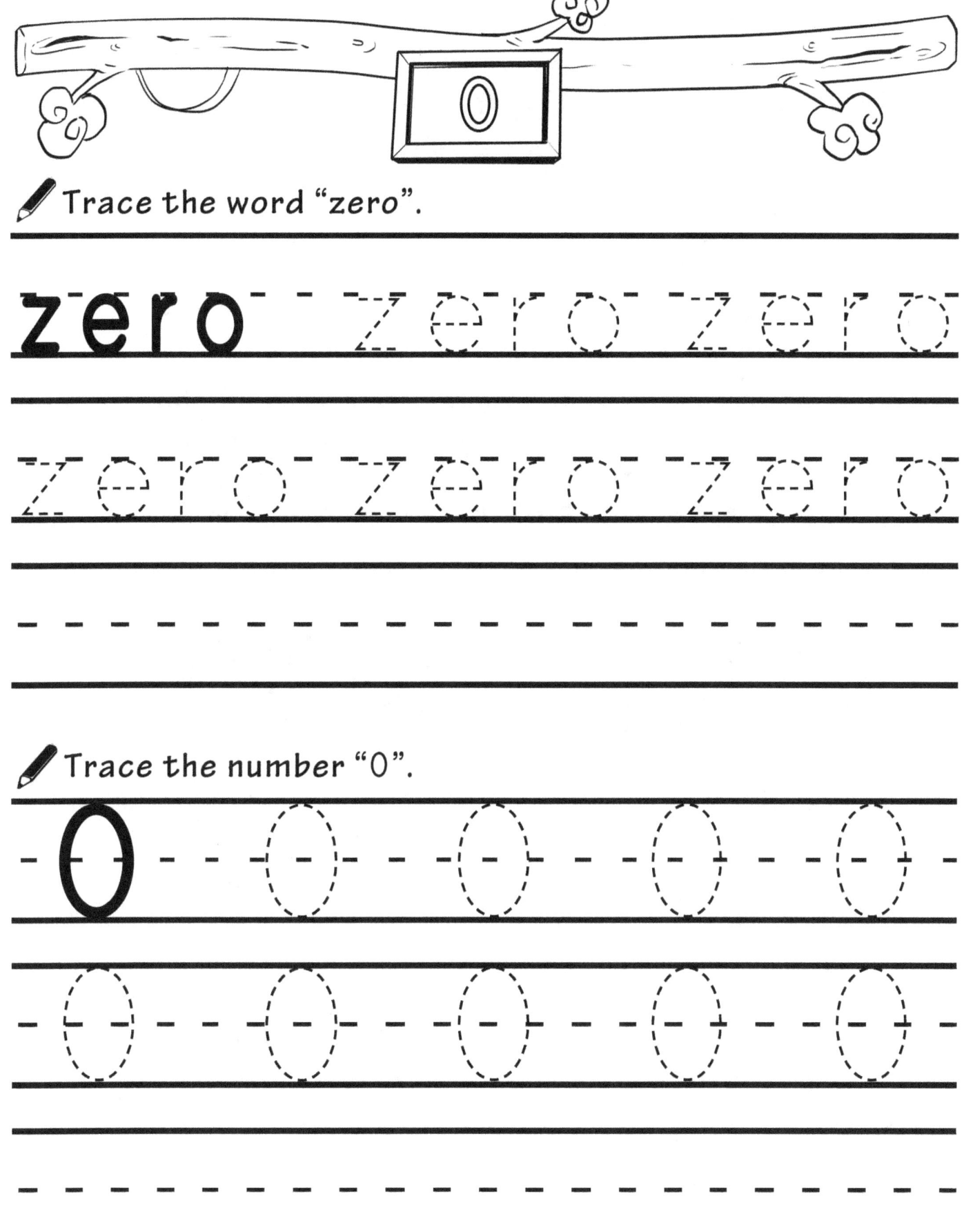

✏️ Trace the word "zero".

zero zero zero
zero zero zero

✏️ Trace the number "0".

0 0 0 0 0
0 0 0 0 0

0 zero

✏ Circle "0" chickens.

✏️ Color the chickens that have "0" dots.

How many fruits are in each box?
Circle your answer.

| | |
|---|---|
| 🍉 | 0 or 1 |
| 🍎🍎 | 2 or 1 |
|  | 1 or 0 |
| 🍐 | 2 or 1 |

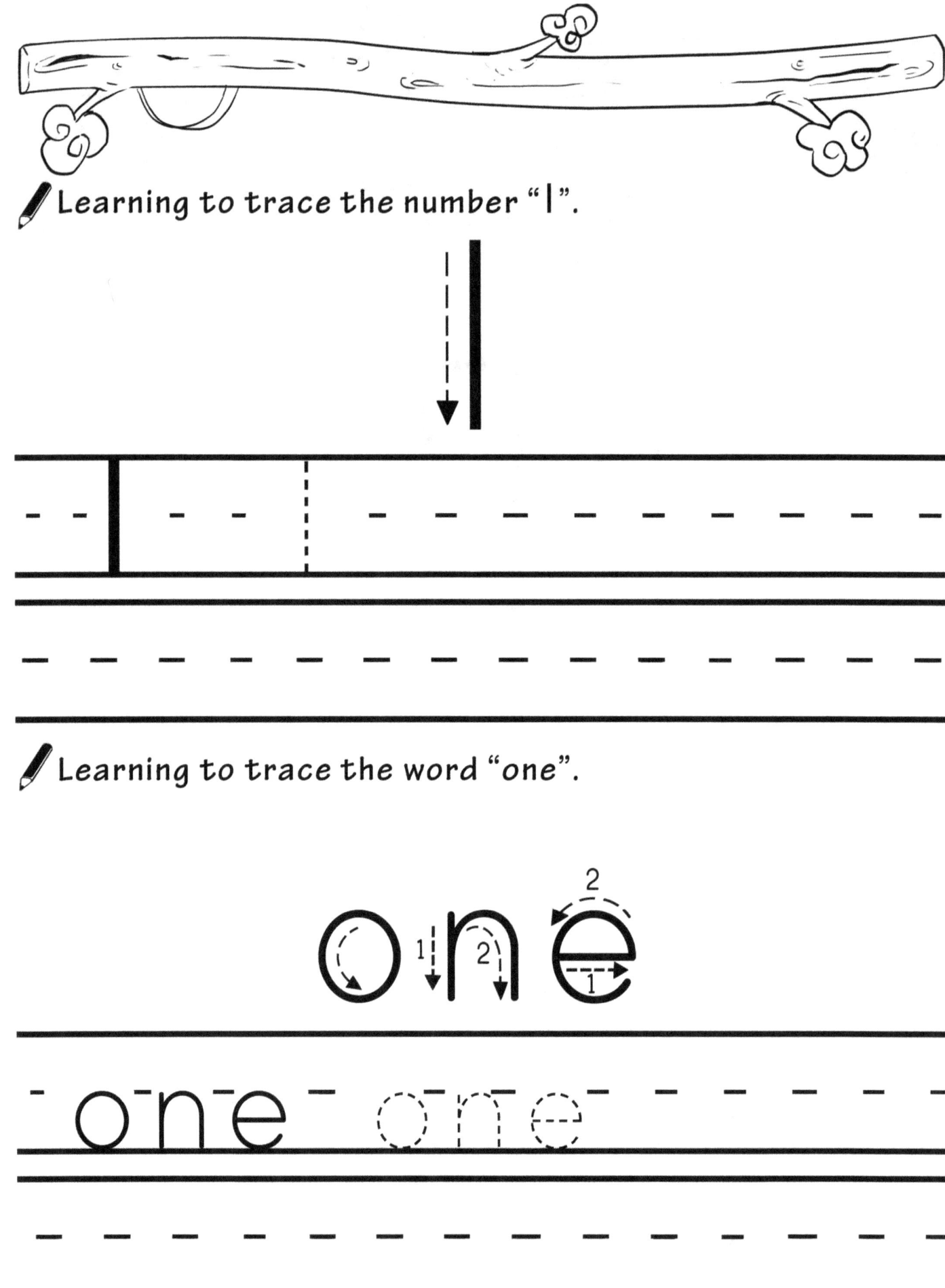

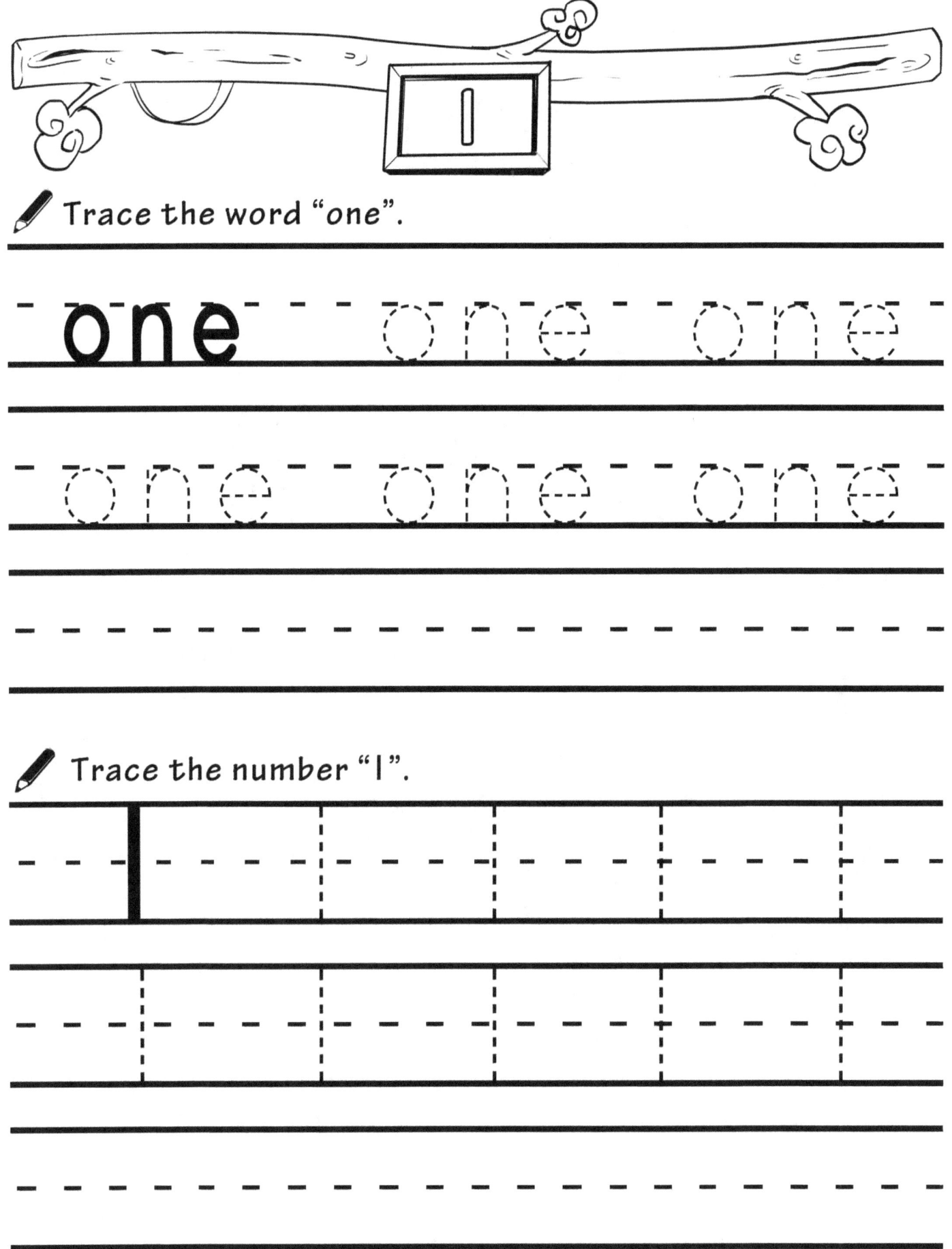

one

# 1 one

✏️ Circle "1" horse.

✏️ Color the horses that have "1" heart.

✏️ Draw a line to the correct number.

2

1

3

✏️ Trace the word "two".

two  two  two
two  two  two

✏️ Trace the number "2".

2  2  2  2
2  2  2  2

two

# 2 two

✏️ Circle "2" pigs.

✏️ Color the pigs that have "2" dots.

## Count and Circle

✏️ **Count and circle the correct number of fruits in each row.**

| 2 | 🍎 🍎 🍎 🍎 |
| 3 | 🍌 🍌 🍌 🍌 |
| 1 | 🍉 🍉 🍉 🍉 |
| 2 | 🍐 🍐 🍐 |
| 0 | 🍓 🍓 🍓 🍓 |

✏️ Learning to trace the number "3".

3  3

✏️ Learning to trace the word "three".

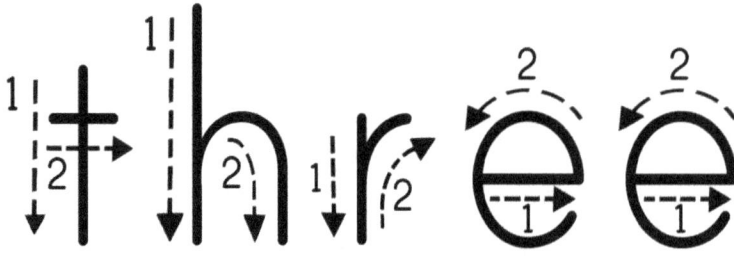

three three

✏️ Trace the word "three".

three three three
three three three

✏️ Trace the number "3".

3 3 3 3 3
3 3 3 3 3

# 3 three

✏️ Circle "**3**" sheep.

✏️ Color the sheep that have "3" flowers.

✏️ Draw 3 fruits on the tree then color it in.

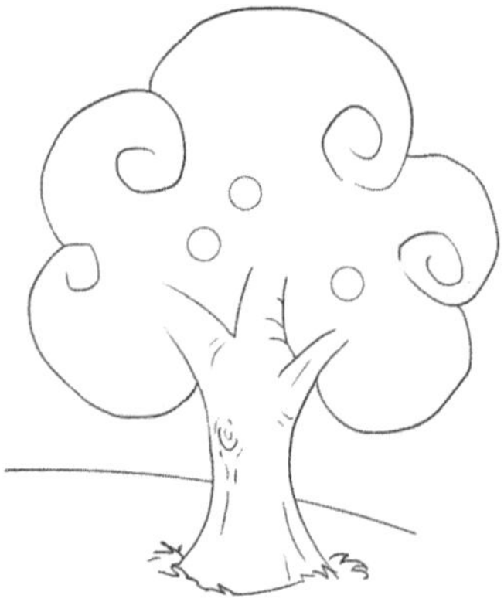

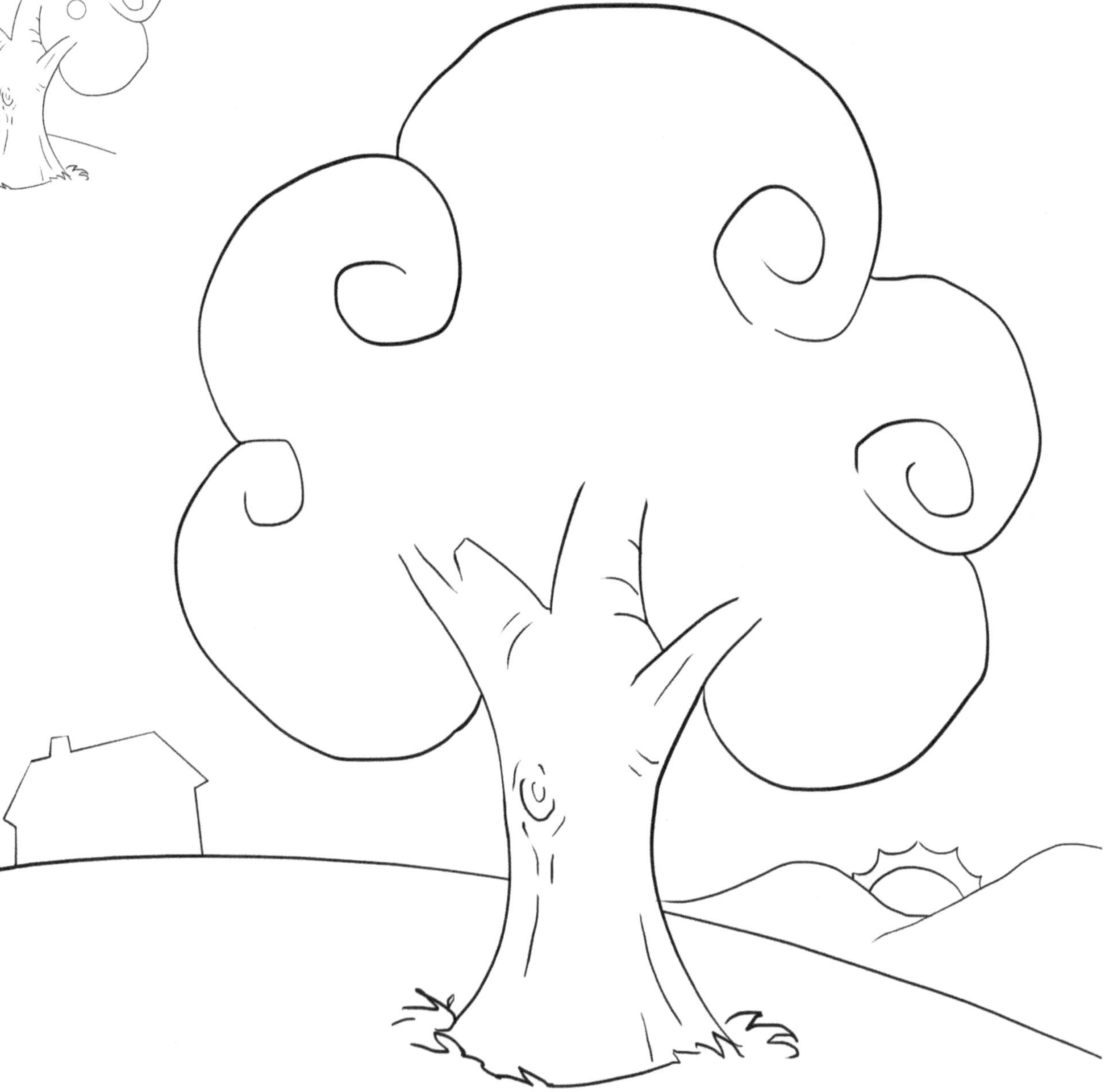

✏️ Learning to trace the number "4".

✏️ Learning to trace the word "four".

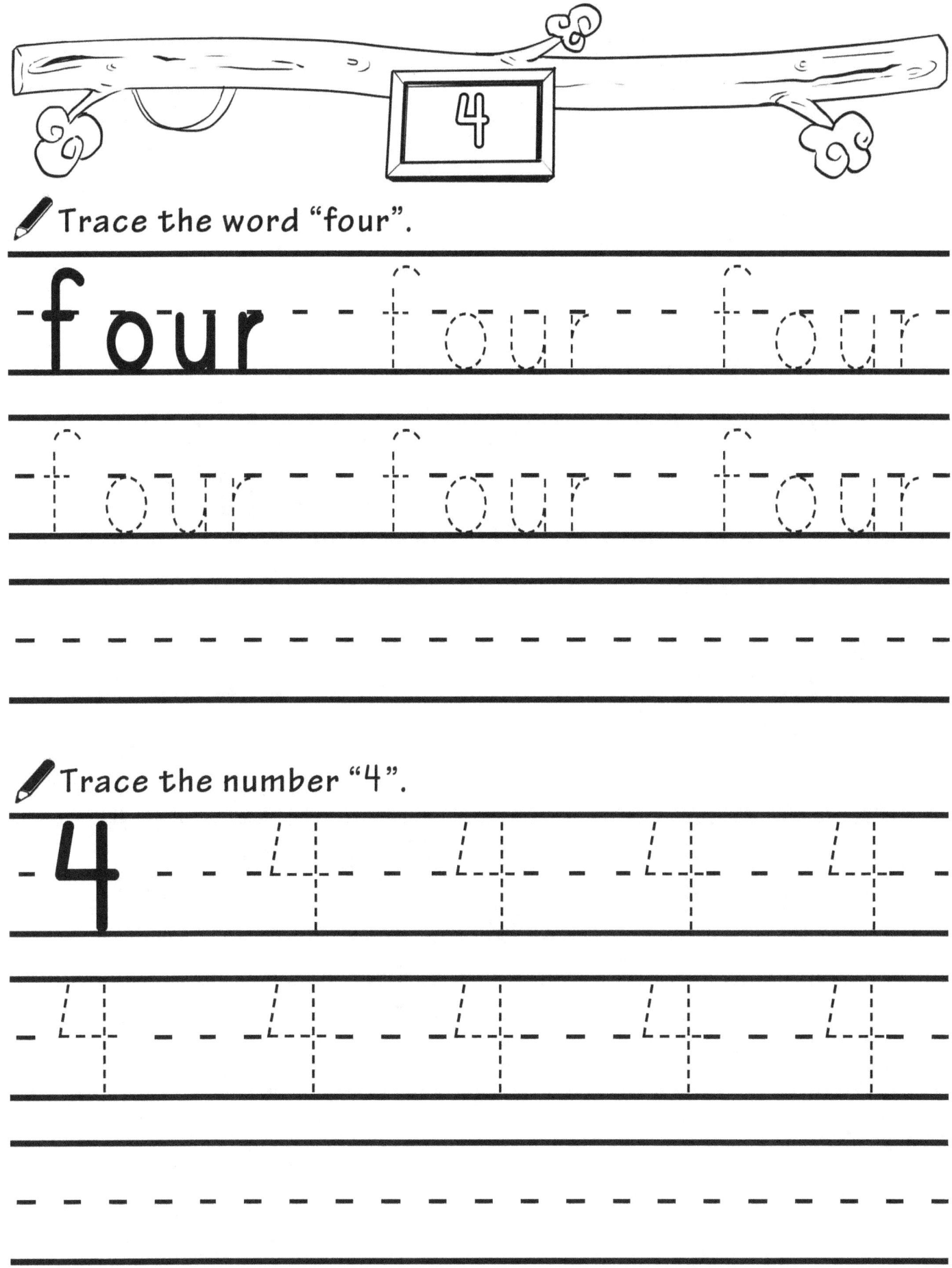

✏️ Trace the word "four".

four  four  four
four  four  four

✏️ Trace the number "4".

4  4  4  4  4
4  4  4  4

four

# 4 four

✏️ Circle "4" cows.

# four

Color the cows that have "4" spots.

✏️ Draw a line to the correct number.

2

1

3

5

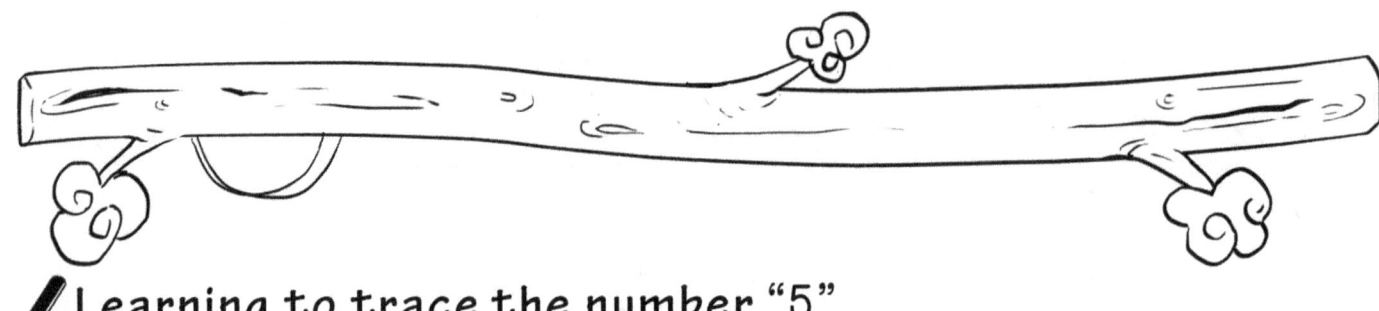

✏️ Learning to trace the number "5".

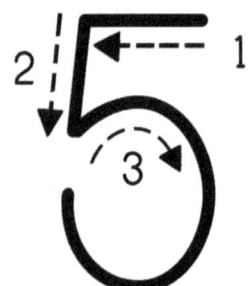

5 5

✏️ Learning to trace the word "five".

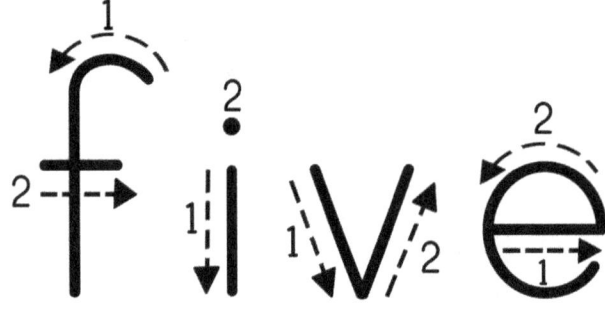

five five

✏️ Trace the word "five".

five   five   five
five   five   five

✏️ Trace the number "5".

5   5   5   5   5
5   5   5   5   5

# 5 five

✏️ Circle "5" goats.

**five**

✏️ Color the goats that have "5" spots.

✏️ Write the numbers below.

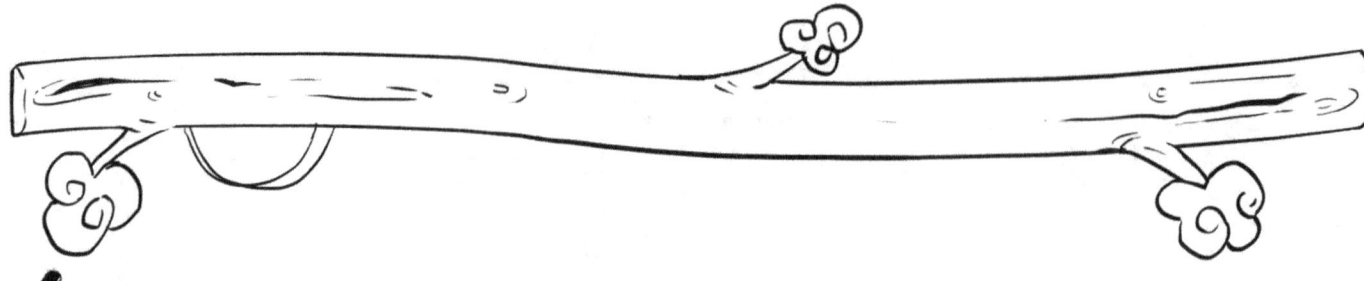

| zero | one | two |
|---|---|---|
| 0 _ _ _ | 1 _ _ _ | 2 _ _ _ |

| three | four | five |
|---|---|---|
| 3 _ _ _ | 4 _ _ _ | 5 _ _ _ |

✏️ Trace the word "six".

✏️ Trace the number "6".

six

# 6 six

✏️ Circle "6" dogs.

# six

✏️ Color the dogs playing with "6" butterflies.

## Draw and Color

✏️ Draw 6 windows on the house then color it in.

✏ Learning to trace the number "7".

✏ Learning to trace the word "seven".

✏️ Trace the word "seven".

**seven** seven seven

seven seven seven

✏️ Trace the number "7".

7 7 7 7 7 7

7 7 7 7 7 7

# 7 seven

✏️ Circle "7" turtles.

✏️ Color the turtles with "7" circles on their shell.

✏️ Draw 7 squares on the turtle's shell then color it in.

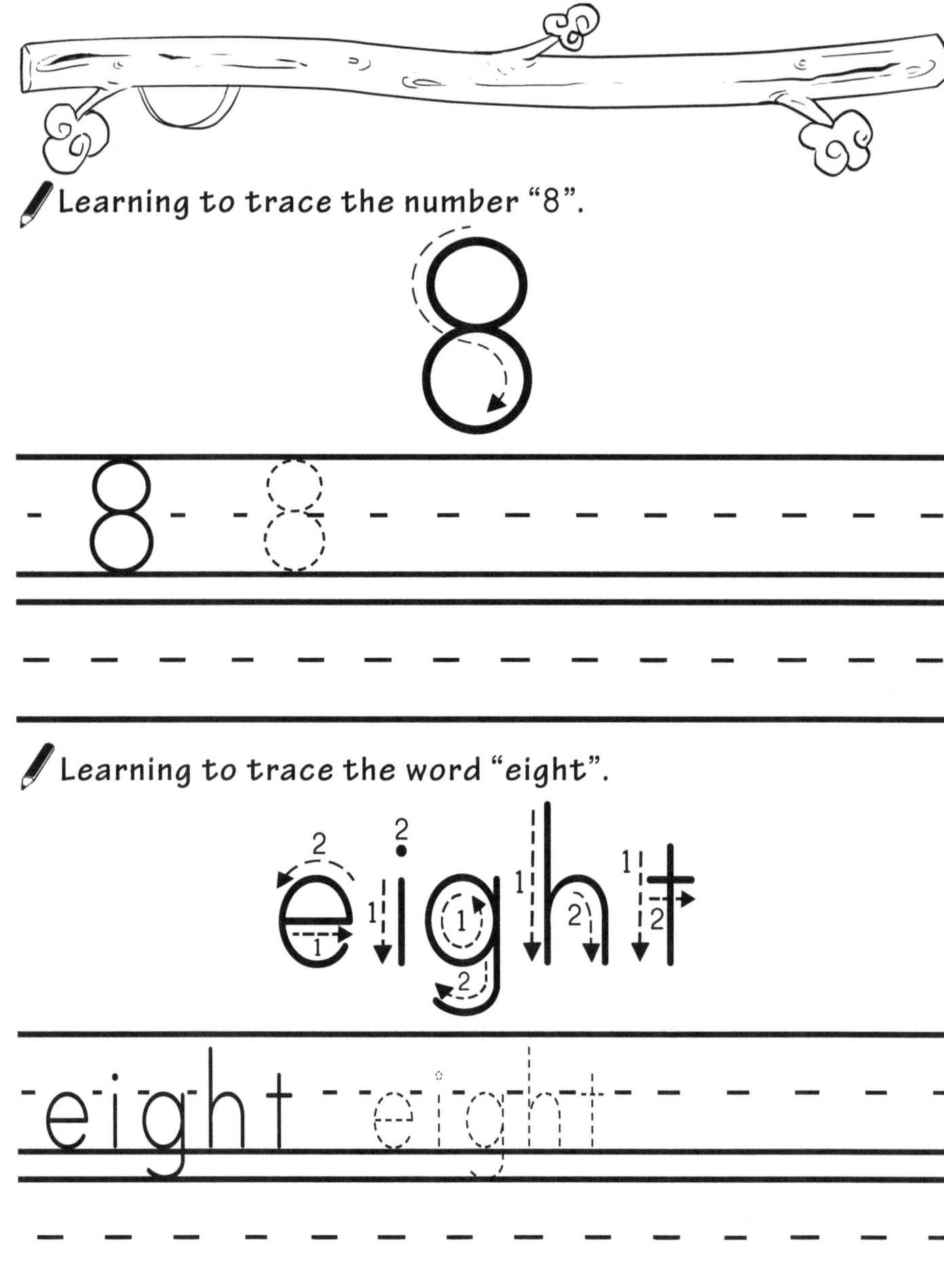

✏️ Learning to trace the number "8".

✏️ Learning to trace the word "eight".

✏️ Trace the word "eight".

eight eight eight
eight eight eight

✏️ Trace the number "8".

8 8 8 8 8
8 8 8 8 8

**eight**

✏️ Color the rabbits playing with "8" bees.

✏️ Draw a line to the correct number.

4

9

6

3

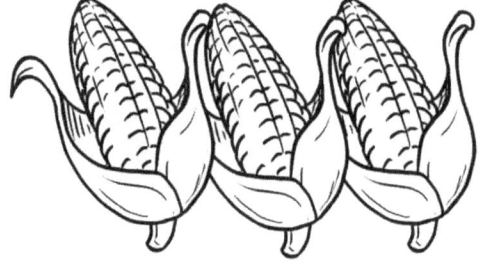

✏️ Learning to trace the number "9".

✏️ Learning to trace the word "nine".

✏ Trace the word "nine".

**nine**   nine   nine

nine   nine   nine

✏ Trace the number "9".

9   9   9   9   9

9   9   9   9   9

nine

# q nine

✏ Circle "9" cats.

✏️ Color the cats playing with "9" stars.

✏️ Draw a line to the correct number.

6

8

10

7

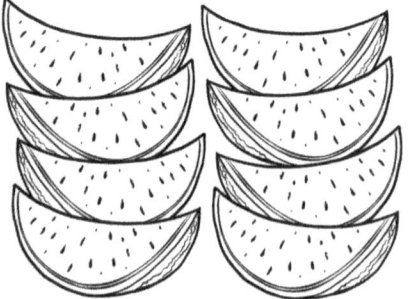

✏️ Learning to trace the number "10".

✏️ Learning to trace the word "ten".

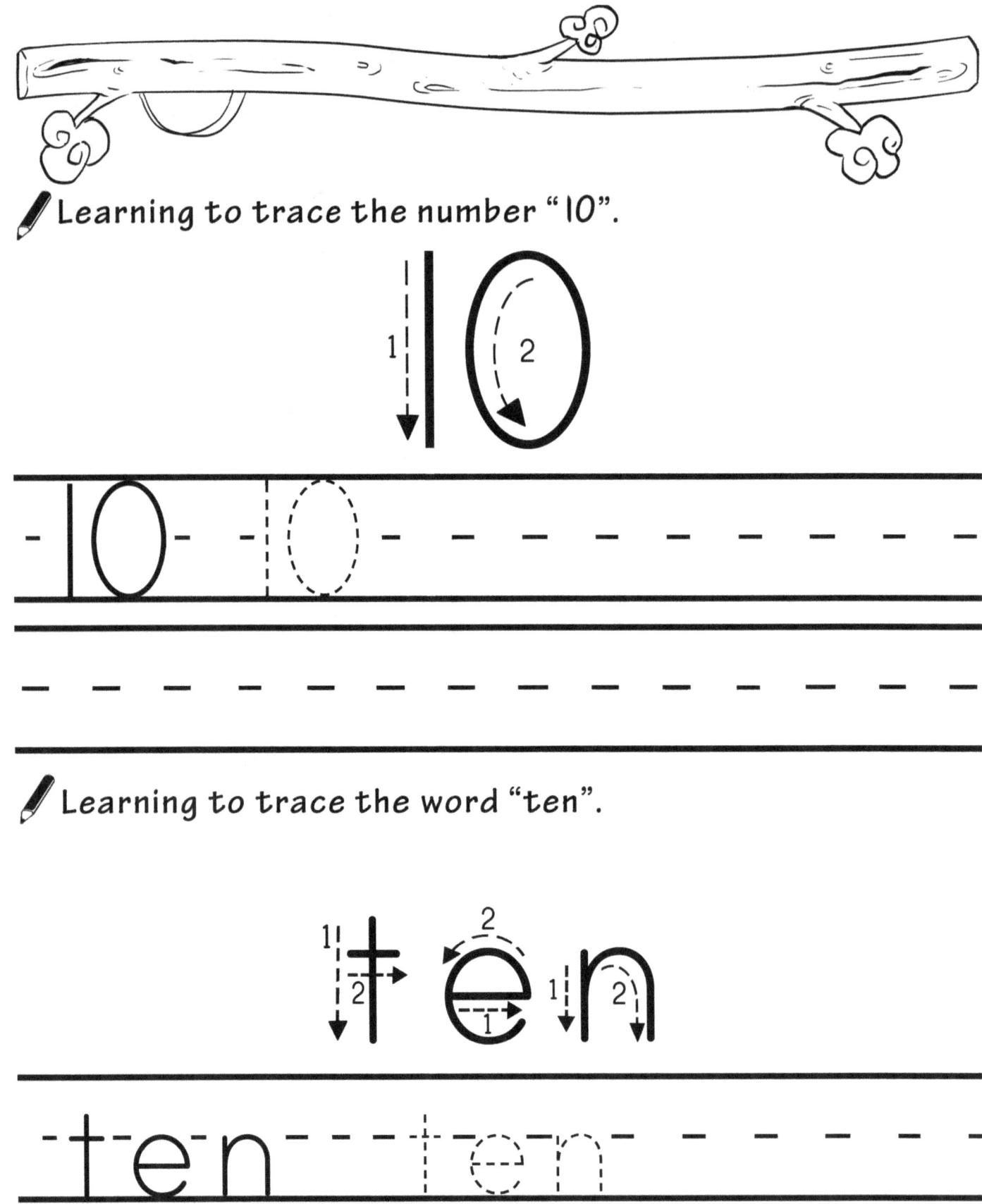

✏️ Trace the word "ten".

ten   ten   ten

ten   ten   ten

✏️ Trace the number "10".

10  10  10  10  10

10  10  10  10  10

ten

# 10 ten

✏️ Circle "10" ducks.

✏️ Color the scene that has "10" ducks.

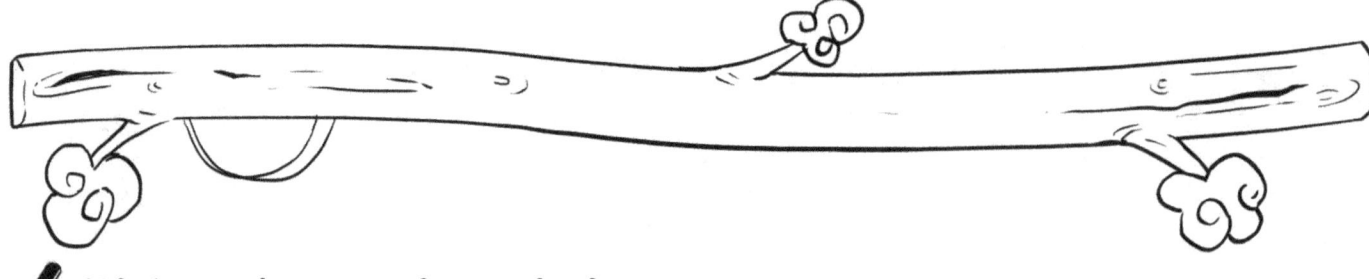

✏️ Write the numbers below.

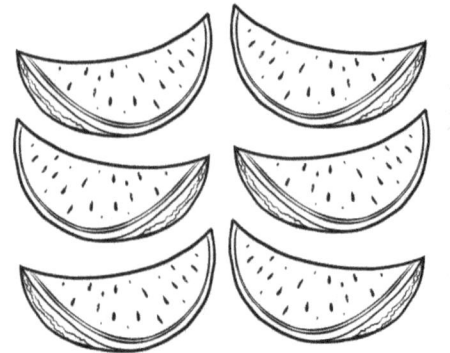

# six

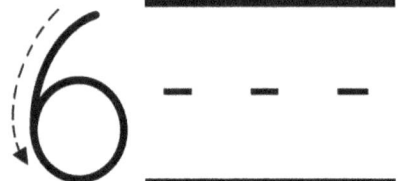

# seven

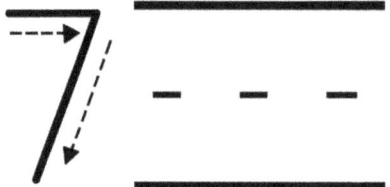

# eight

# nine

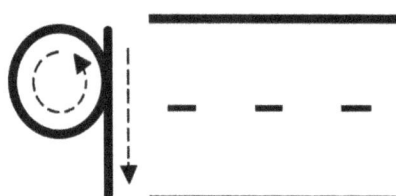

# ten

The number 1 comes "before" numbers 2 and 3

__1__ 2 3

✏️ Write the number that comes "before" these numbers.

| ___ 5 6 | ___ 1 2 |
| ___ 2 3 | ___ 3 4 |
| ___ 8 9 | ___ 4 5 |

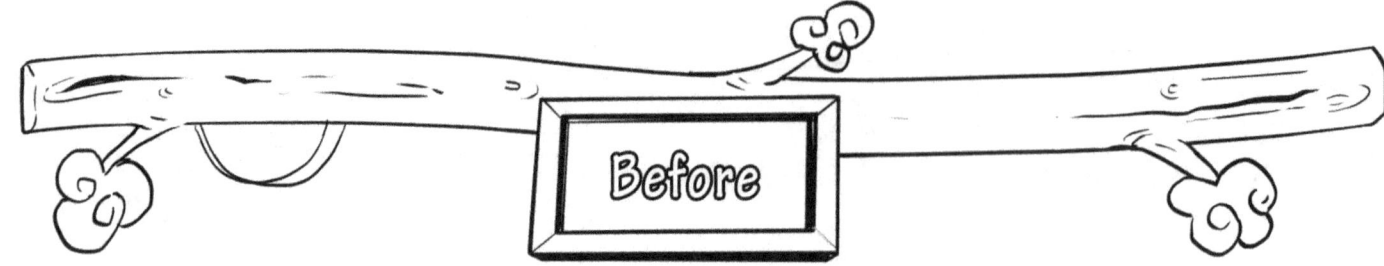

**Before**

The number 1 comes "*before*" numbers 2 and 3

1  2  3

✏ Write the number that comes "*before*" these numbers.

| ___ 6 7 | ___ 3 4 |
| ___ 4 5 | ___ 9 10 |
| ___ 7 8 | ___ 5 6 |

The number 1 comes "*before*" numbers 2 and 3

**1 2 3**

✏️ Write the number that comes "*before*" these numbers.

| ___ 2 3 | ___ 8 9 |
|---|---|
| ___ 6 7 | ___ 7 8 |
| ___ 1 2 | ___ 9 10 |

The number 4 comes "after" numbers 2 and 3

2 3 4

✏️ Write the number that comes "after" these numbers.

| | |
|---|---|
| 5 6 ___ | 1 2 ___ |
| 2 3 ___ | 8 9 ___ |
| 6 7 ___ | 4 5 ___ |

### After

The number 4 comes "after" numbers 2 and 3

2  3  4

✏️ Write the number that comes "after" these numbers.

| | |
|---|---|
| 7 8 ___ | 5 6 ___ |
| 0 1 ___ | 6 7 ___ |
| 1 2 ___ | 3 4 ___ |

The number 4 comes "after" numbers 2 and 3

2　3　4

✏️ Write the number that comes "after" these numbers.

| 2 3 ___ | 8 9 ___ |
|---|---|
| 4 5 ___ | 3 4 ___ |
| 0 1 ___ | 7 8 ___ |

## Color the Scene

✏️ Use the color key to color the scene.

⑤ YELLOW   ⑥ BROWN   ⑦ GREEN

**Between**

The number **2** is in "*between*" numbers **1** and **3**

1 — 2 — 3

✏️ Write the number that is in "*between*" these numbers.

| 4 — — — 6 | 2 — — — 4 |
|---|---|
| 7 — — — 9 | 8 — — — 10 |
| 1 — — — 3 | 6 — — — 8 |

The number 2 is in "*between*" numbers 1 and 3

1 — 2 — 3

✏ Write the number that is in "*between*" these numbers.

| 5 — — 7 | 3 — — 5 |
|---|---|
| 1 — — 3 | 4 — — 6 |
| 7 — — 9 | 0 — — 2 |

The number 2 is in "*between*" numbers 1 and 3

1 2 3

✏ Write the number that is in "*between*" these numbers.

| 0 _ _ _ 2 | 6 _ _ _ 8 |
|---|---|
| 2 _ _ _ 4 | 5 _ _ _ 7 |
| 8 _ _ _ 10 | 3 _ _ _ 5 |

# Left or Right

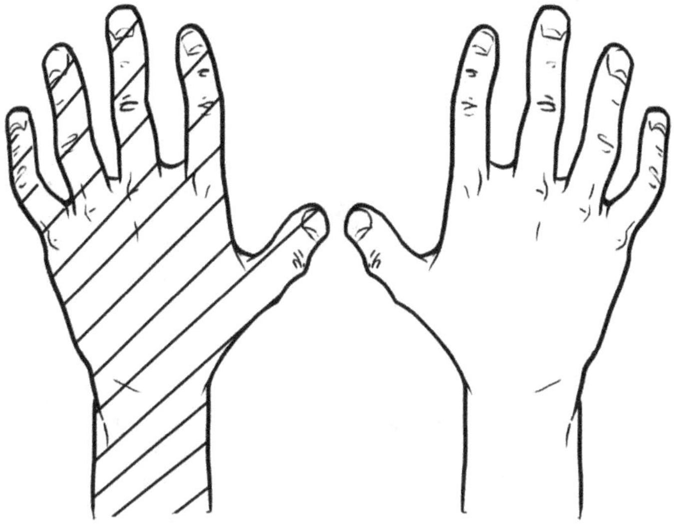

| Left | Right | Left | Right |

🖉 We have shaded the hand on the left. | 🖉 We have shaded the hand on the right.

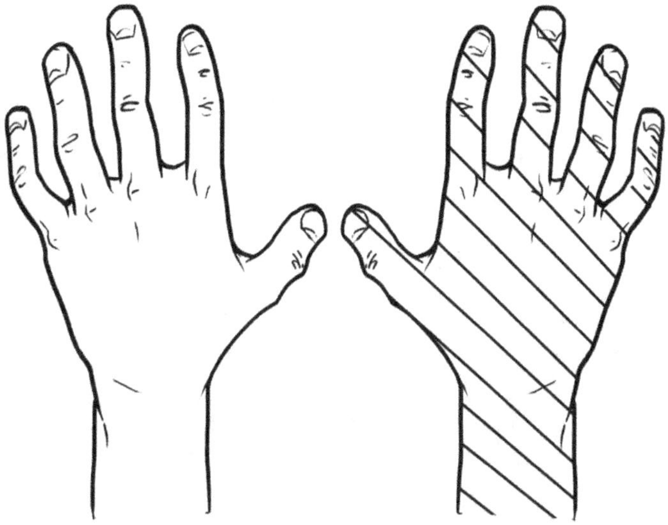

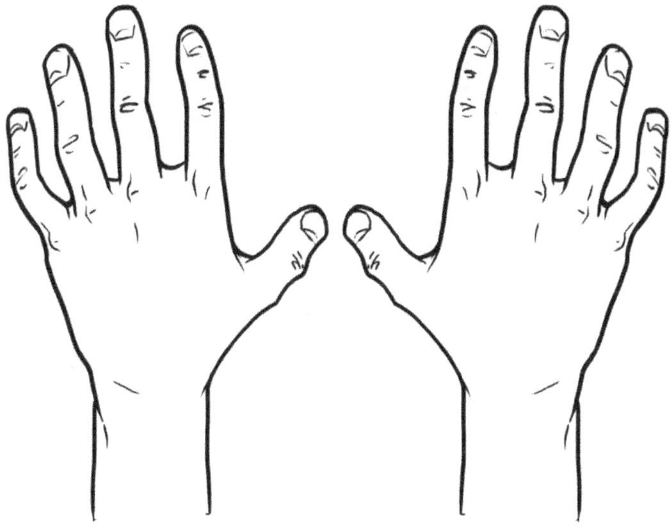

🖉 Color the hand on the left. 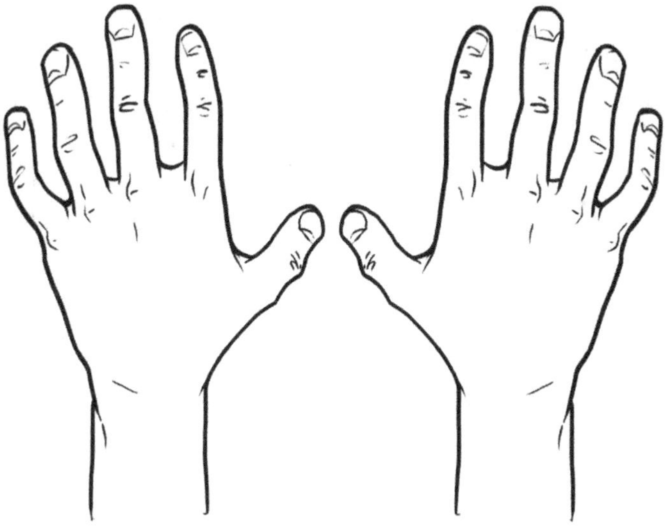 | 🖉 Color the hand on the right.

# Left or Right

Color the animal on the left.

Color the animal on the right.

Color the fruits on the right.

Color the vegetables on the left.

# Left or Right

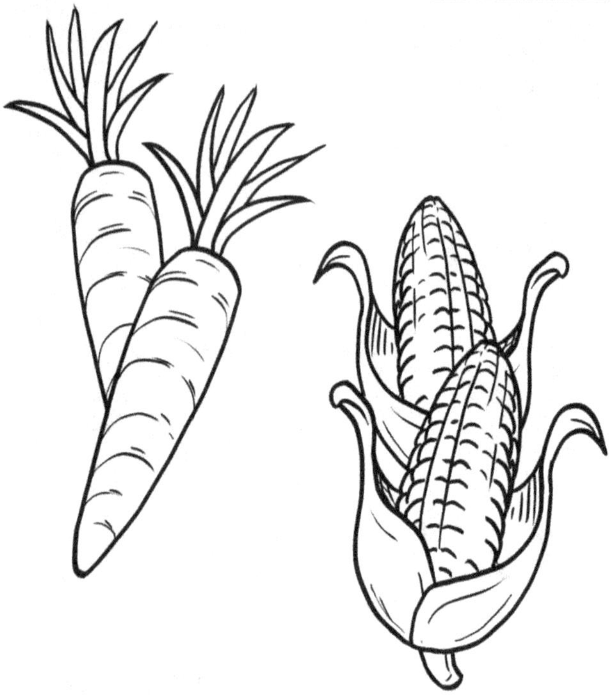

Color the vegetables on the left.

Color the fruits on the right.

Color the fruits on the right.

Color the animal on the right.

# Top or Bottom

Color the animal at the bottom.

Color the fruits at the top.

Color the fruits at the top.

Color the animals at the bottom.

How many animals are in each box?
Circle your answer.

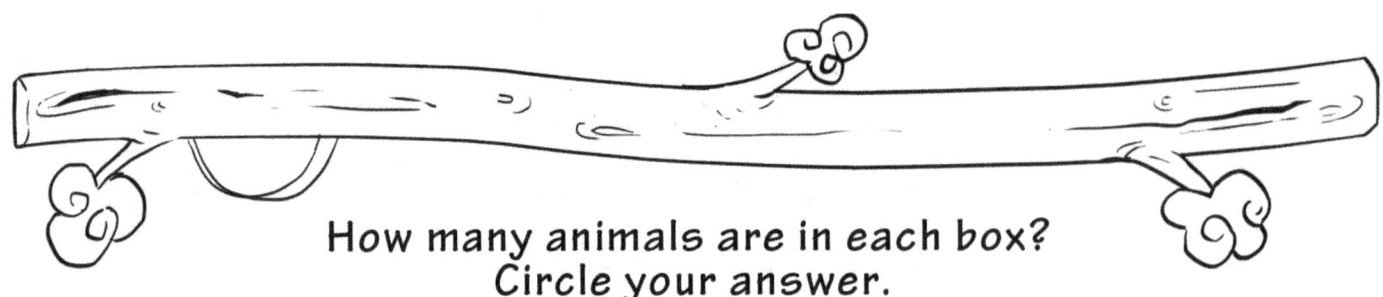

9 or 10

4 or 5

6 or 7

3 or 4

How many vegetables are in each box?
Circle your answer.

| | |
|---|---|
| 🎃 pumpkins (5) | 4 or 5 |
| 🥕 carrots (8) | 7 or 8 |
| 🌽 corn (5) | 5 or 6 |
| 🧅 onions (8) | 8 or 9 |

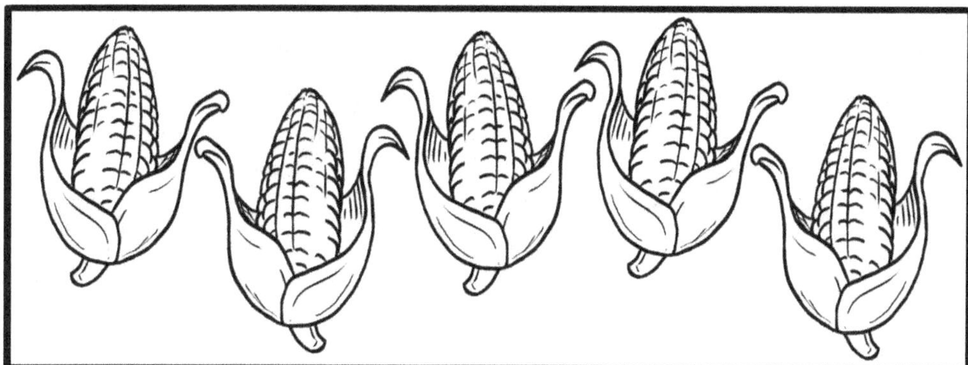

How many fruits are in each box?
Circle your answer.

| | |
|---|---|
| (4 watermelon slices) | 3 or 4 |
| (6 apples) | 6 or 7 |
| (9 strawberries) | 8 or 9 |
| (3 pears) | 2 or 3 |

How many animals are in each box?
Circle your answer.

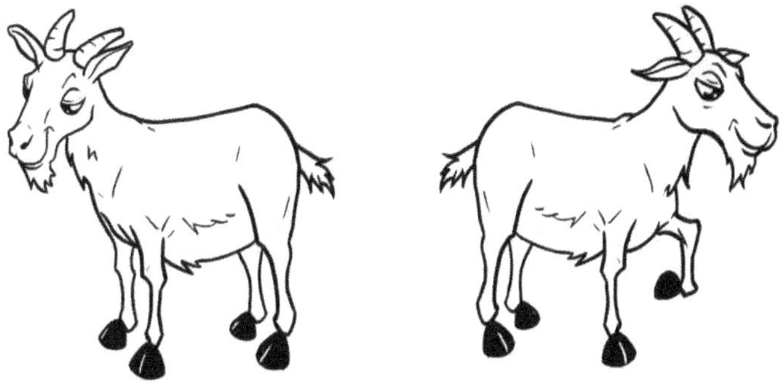

| | |
|---|---|
| (two goats) | 2 or 3 |
| (eight ducks) | 7 or 8 |
| (two sheep) | 1 or 2 |

3 or 4

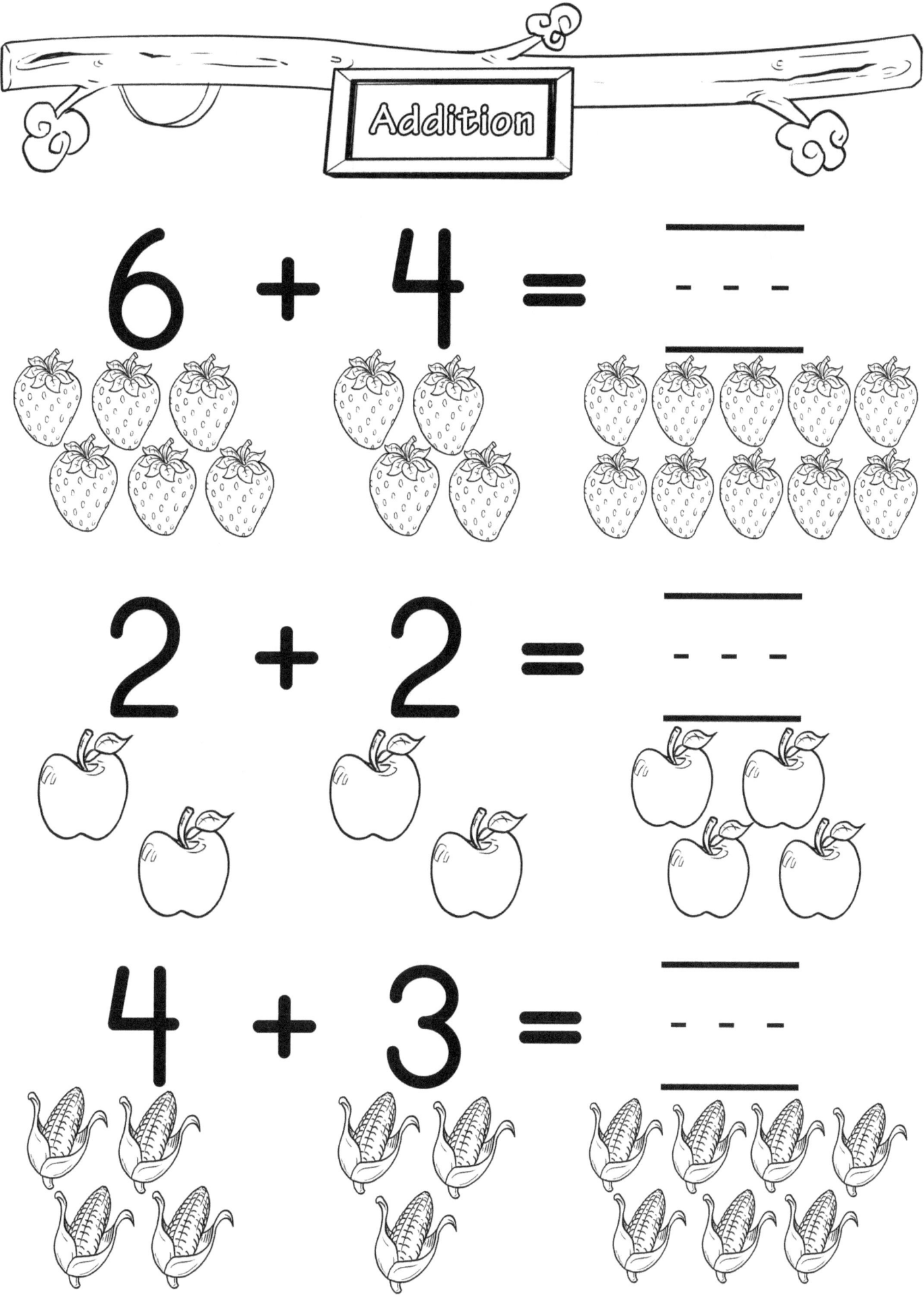

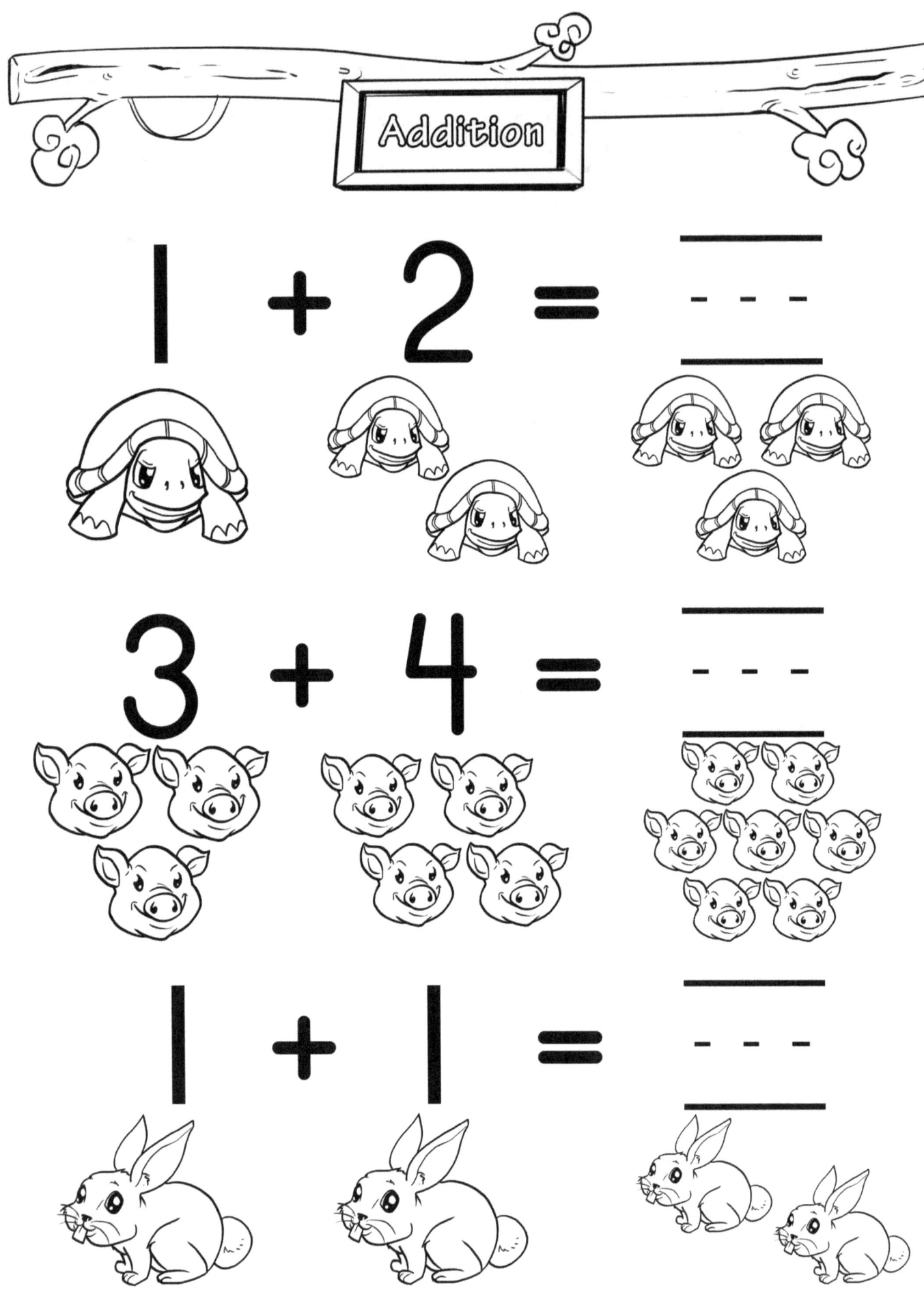

✏ Connect the dots from number 1 to 10.

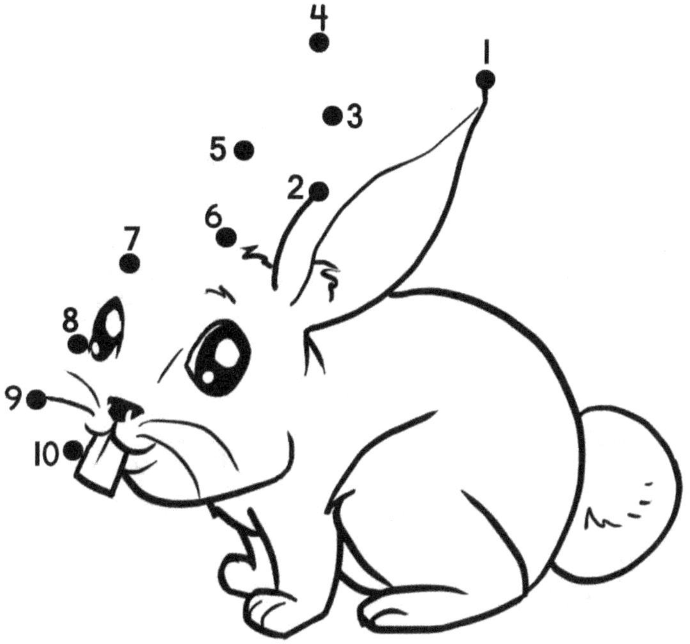

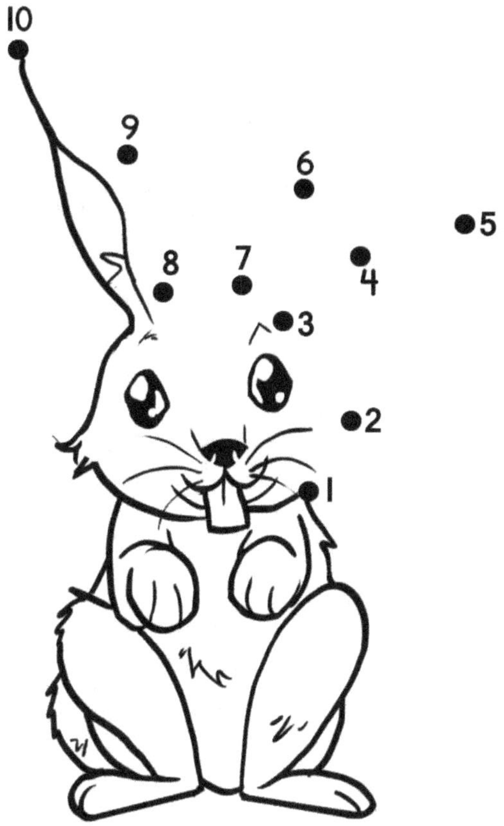

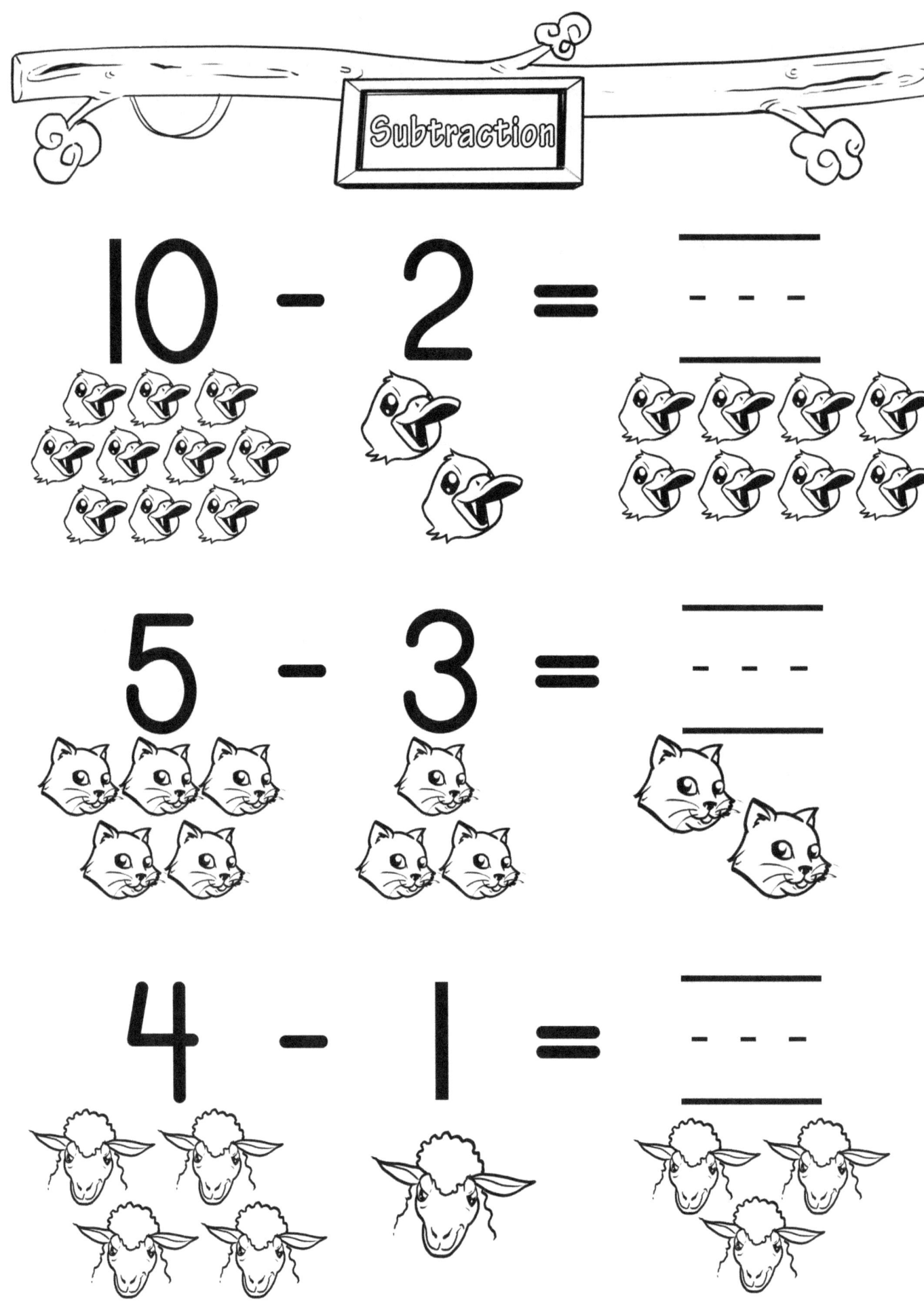

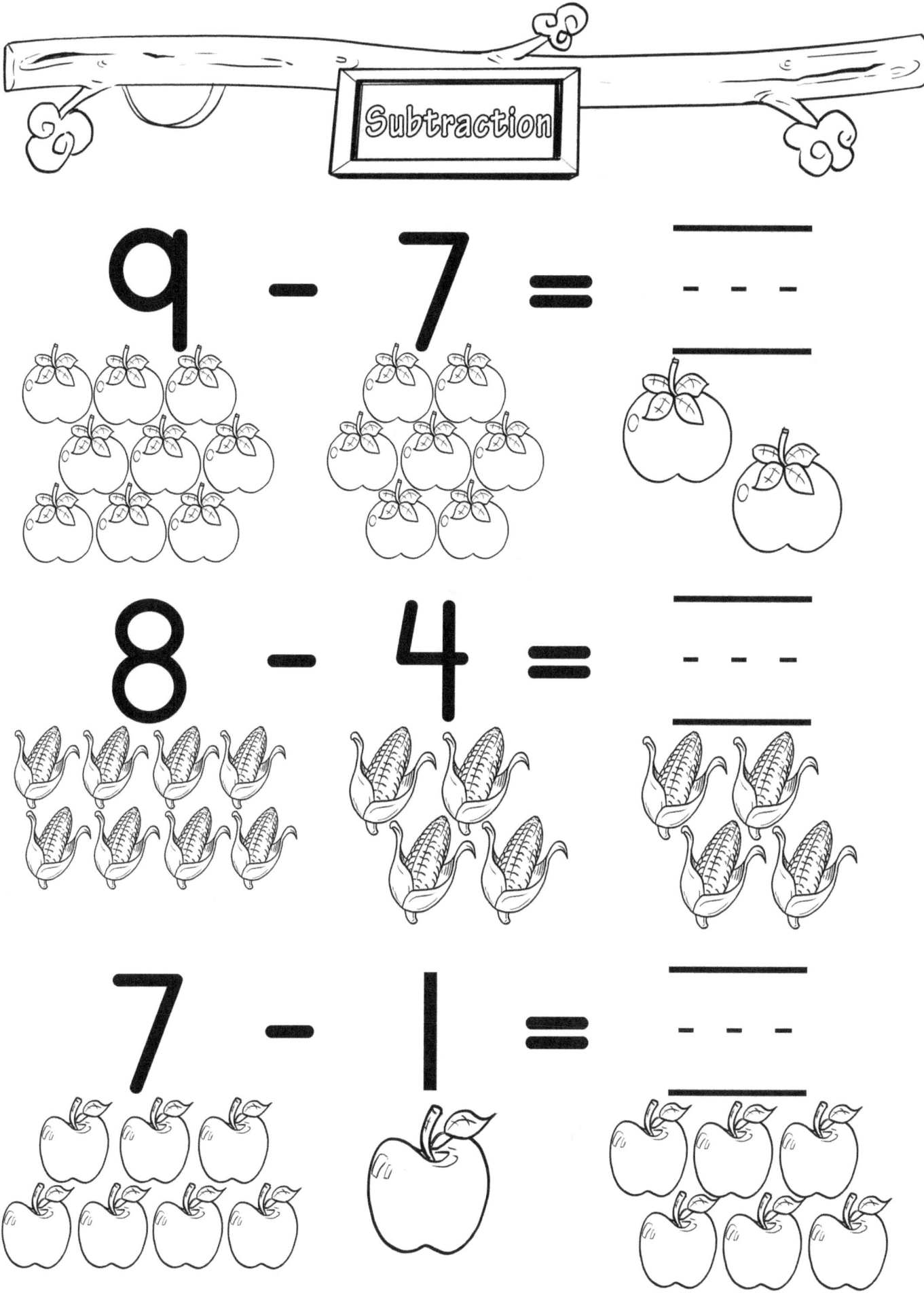

✏️ Connect the dots from number 1 to 10.

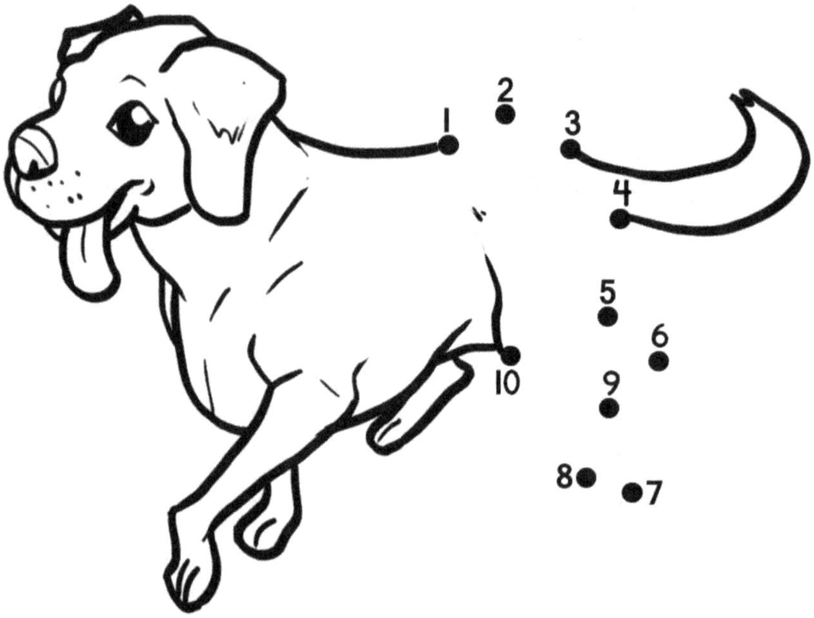

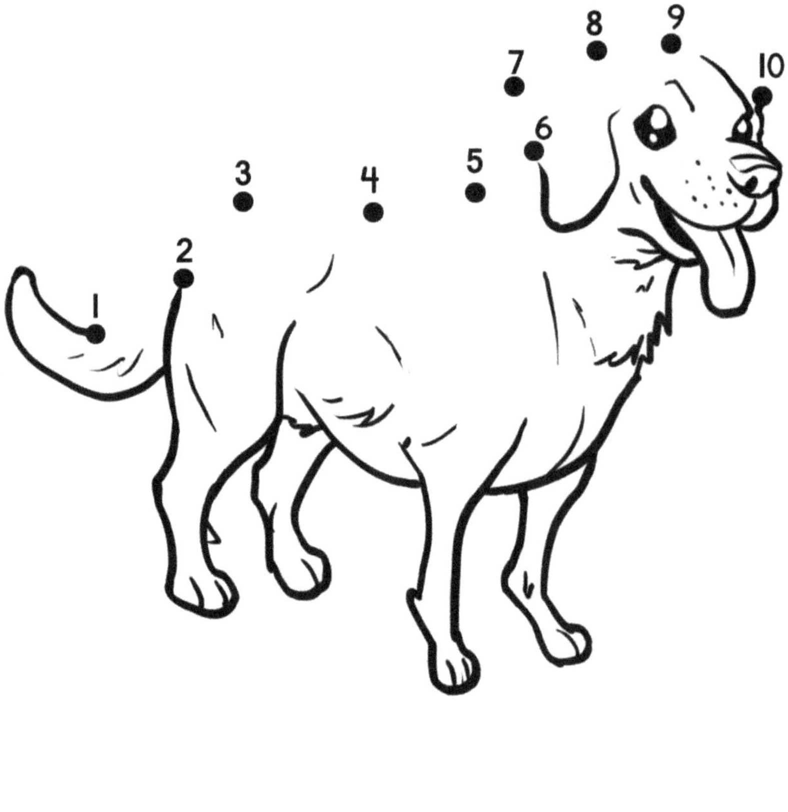

✏️ Count and circle the correct number of animals in each row.

| | |
|---|---|
| 9 | (birds) |
| 5 | (rabbits) |
| 3 | (cats) |
| 1 | (cows) |
| 2 | (horses) |

## Count and Circle

✏️ Count and circle the correct number of fruits in each row.

| 10 | 🍎🍎🍎🍎🍎🍎🍎🍎🍎🍎 |
| 7 | 🍌🍌🍌🍌🍌🍌🍌 |
| 1 | 🍉🍉🍉🍉 |
| 6 | 🍐🍐🍐🍐🍐🍐🍐 |
| 4 | 🍓🍓🍓🍓🍓🍓 |

✏️ Count and circle the correct number of vegetables in each row.

| 1 | 🎃 🎃 |
|---|---|
| 3 | 🌽 🌽 🌽 🌽 🌽 |
| 7 | 🍎 🍎 🍎 🍎 🍎 🍎 🍎 |
| 5 | 🥕 🥕 🥕 🥕 🥕 |

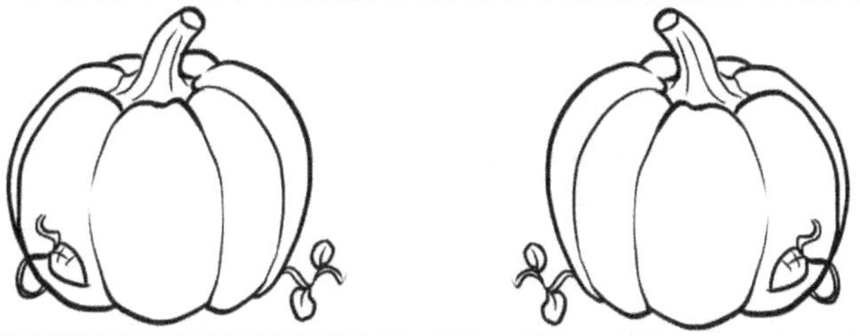

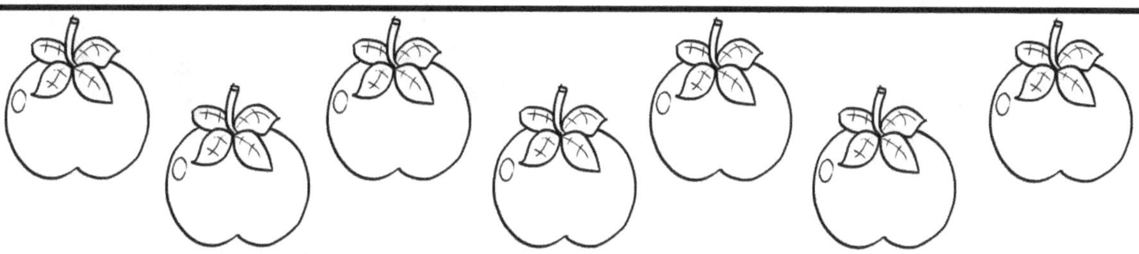

## Connect the Dots

✏️ Connect the dots from number 1 to 10.

# Shapes

✏️ Trace, complete and draw the shapes.

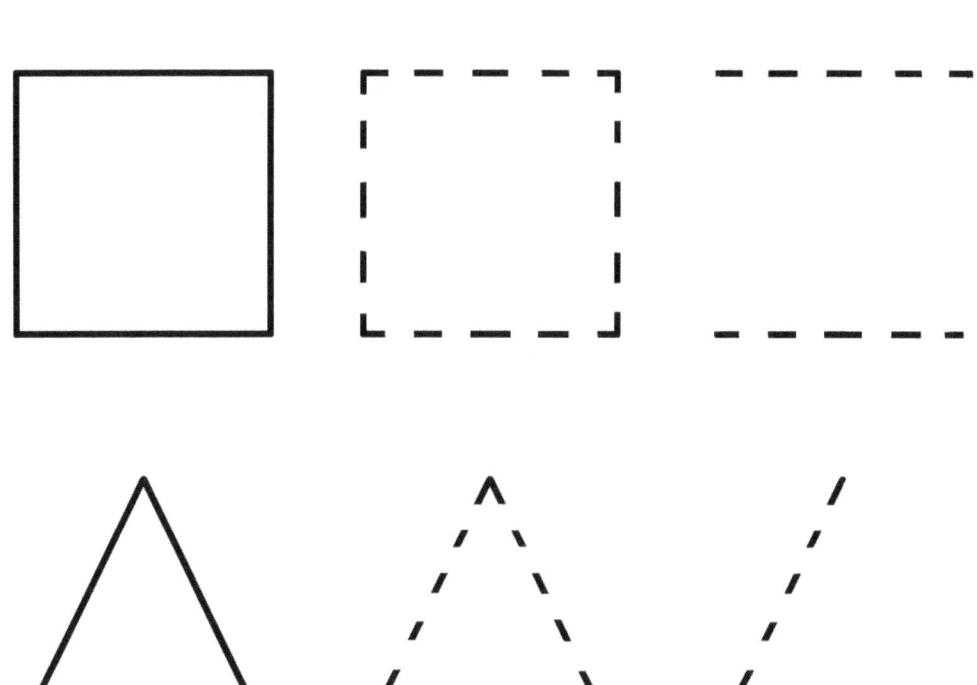

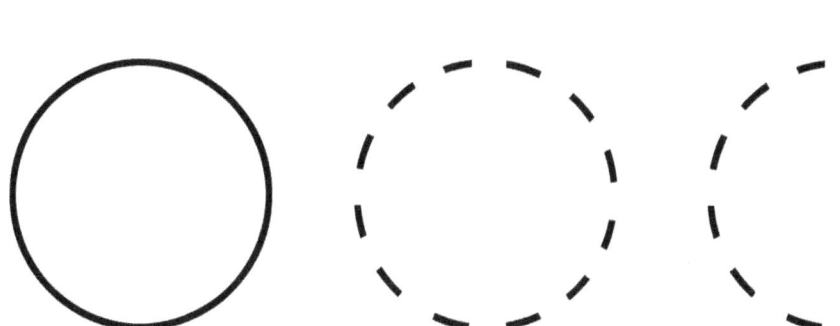

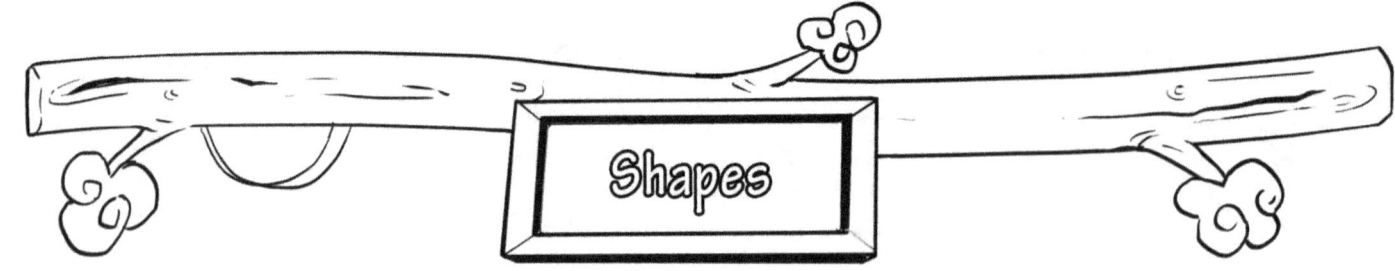

✏️ Write how many sides each shape has.

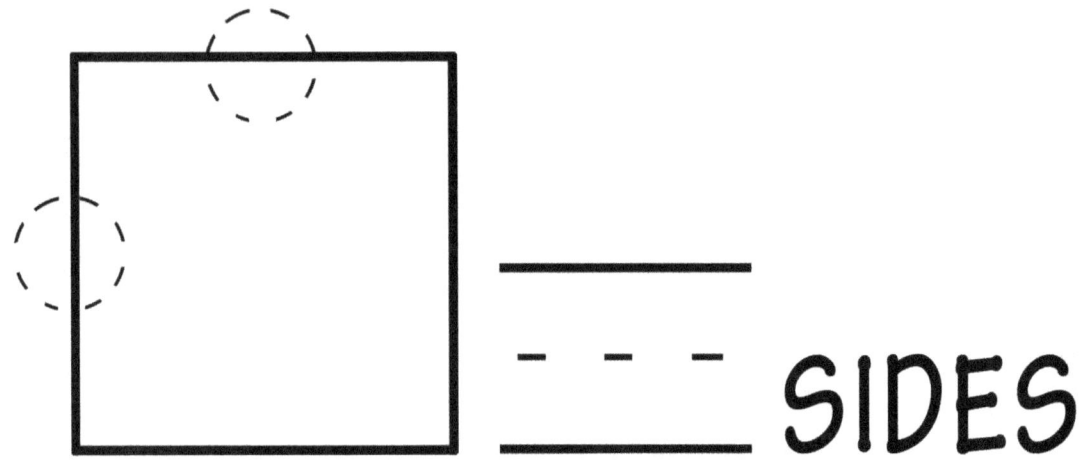

___
___ SIDES

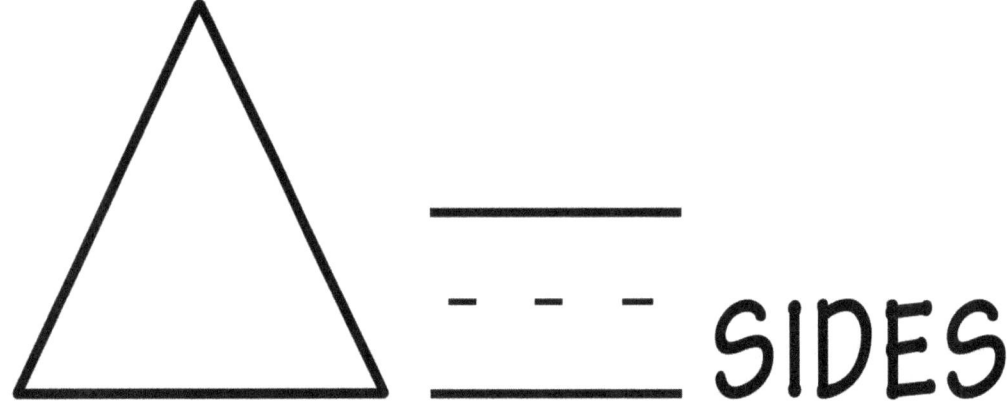

___
___ SIDES

✏ Write how many corners each shape has.

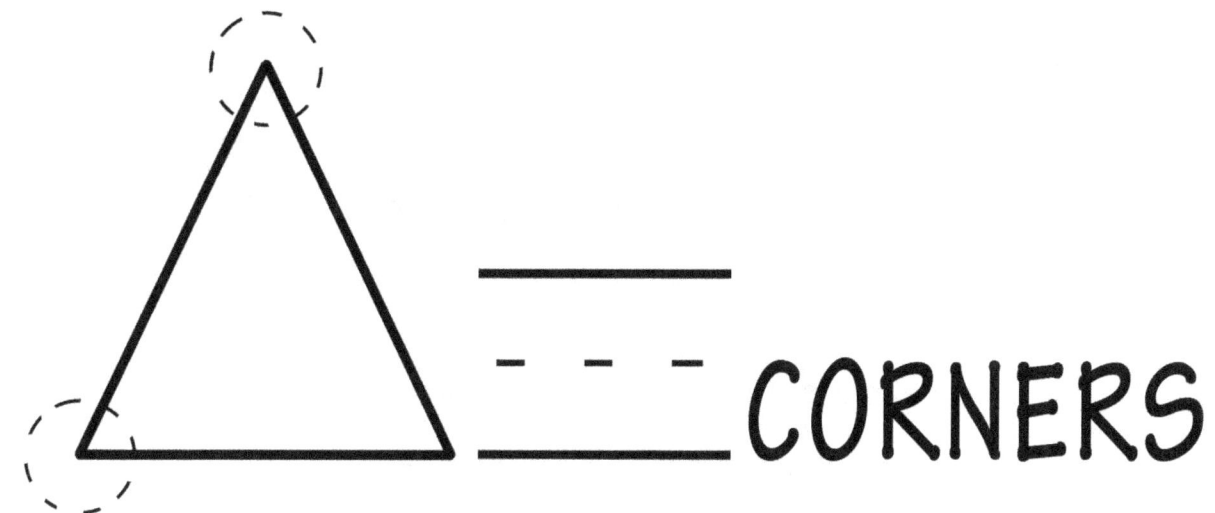

_ _ _ _ CORNERS

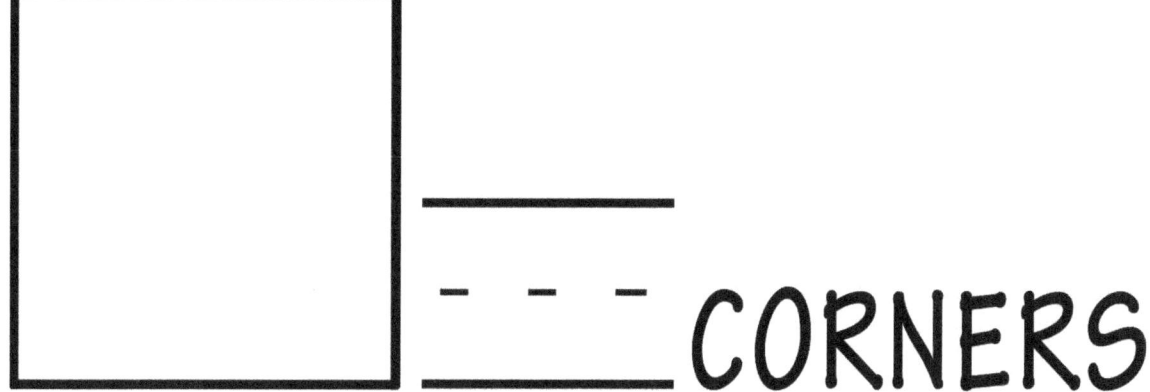

_ _ _ _ CORNERS

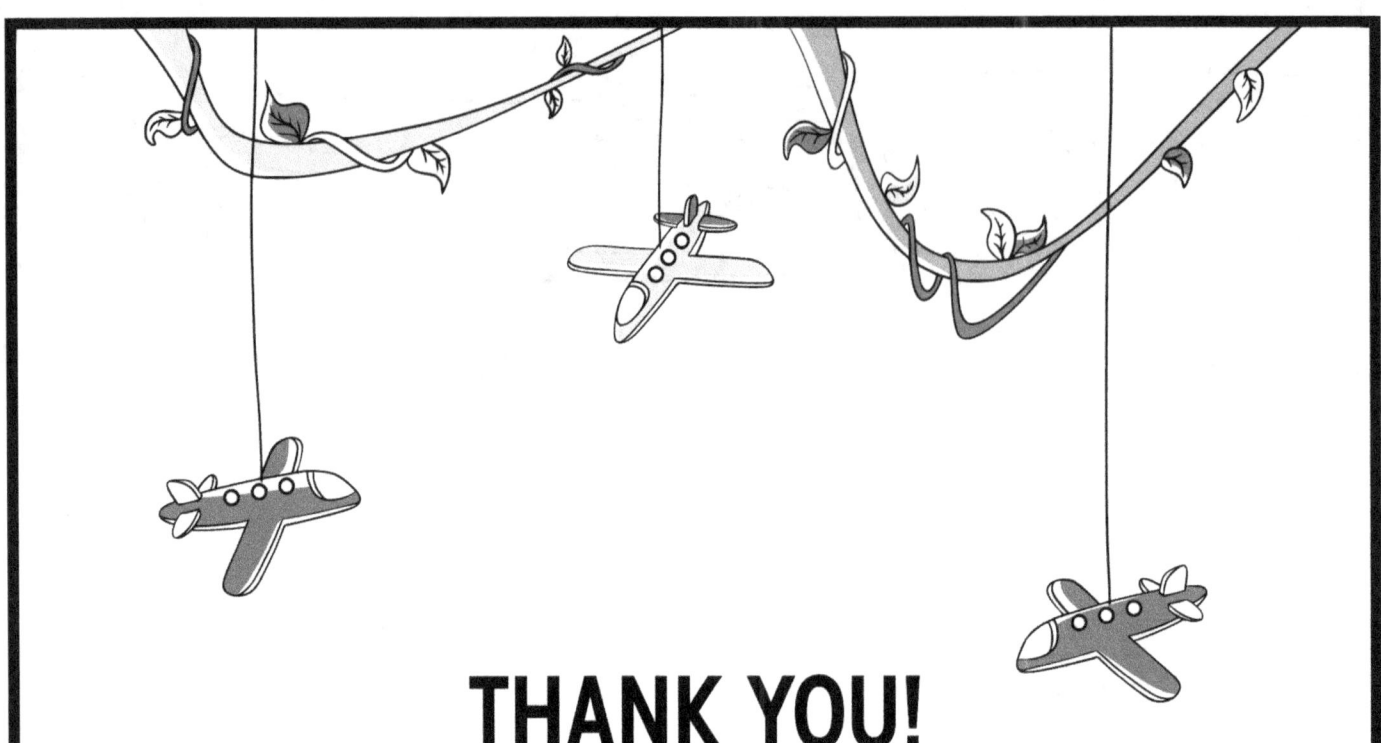

# THANK YOU!

We hope you enjoyed the book.
Please consider leaving a review
where you bought it!

For more, please visit:
wizolearning.com